EXPLORE BIOMES

SAVANNA BIOMES

BY CLARA MacCARALD

Kids Core

An Imprint of Abdo Publishing
abdobooks.com

abdobooks.com

Published by Abdo Publishing, a division of ABDO, PO Box 398166, Minneapolis, Minnesota 55439. Copyright © 2024 by Abdo Consulting Group, Inc. International copyrights reserved in all countries. No part of this book may be reproduced in any form without written permission from the publisher. Kids Core™ is a trademark and logo of Abdo Publishing.

Printed in the United States of America, North Mankato, Minnesota.
052023
092023

THIS BOOK CONTAINS RECYCLED MATERIALS

Cover Photo: Shutterstock Images
Interior Photos: Volodymyr Burdiak/Shutterstock Images, 4–5, 25; Joe McDonald/The Image Bank/Getty Images, 6; Eray Bozkurt/Shutterstock Images, 8; Valentin Wolf/Image Broker/Getty Images, 10–11; Shutterstock Images, 12, 16–17, 26, 28 (ostrich), 29 (giraffe, zebra); Anup Shah/Nature Picture Library/Alamy, 14; Gudkov Andrey/Shutterstock Images, 19; Martin Prochazkacz/Shutterstock Images, 20; Jonathan C Photography/Shutterstock Images, 22; Daria Riabets/Shutterstock Images, 28 (lion); Blue Ring Media/Shutterstock Images, 28–29 (background)

Editor: Angela Lim
Series Designer: Ryan Gale

Library of Congress Control Number: 2022949089

Publisher's Cataloging-in-Publication Data

Names: MacCarald, Clara, author.
Title: Savanna biomes / by Clara MacCarald
Description: Minneapolis, Minnesota: Abdo Publishing Company, 2024 | Series: Explore biomes | Includes online resources and index.
Identifiers: ISBN 9781098291136 (lib. bdg.) | ISBN 9781098277314 (ebook)
Subjects: LCSH: Savannas--Juvenile literature. | Biotic communities--Juvenile literature. | Habitats--Juvenile literature. | Life zones--Juvenile literature. | Savanna animals--Juvenile literature. | Savanna plants--Juvenile literature. | Savanna ecology--Juvenile literature.
Classification: DDC 577--dc23

CONTENTS

CHAPTER 1
A Savanna Hunt 4

CHAPTER 2
About Savannas 10

CHAPTER 3
Animals and Plants in Savannas 16

Explore the Biome 28
Glossary 30
Online Resources 31
Learn More 31
Index 32
About the Author 32

Living in a herd protects zebras from the animals that hunt them.

A SAVANNA HUNT

The sun dips low as evening approaches on the African savanna. A herd of zebras strolls along. Yellow and green grasses reach as high as the animals' bellies. A few trees are scattered across the plain.

Lions typically eat large animals that weigh more than 100 pounds (45 kg).

Grass rustles nearby. The zebras look toward the noise, alarmed. An African lion leaps out. The big cat's yellow fur made it difficult to see among the plants. The zebras run as a herd. More lions appear from other directions. They all belong to the same group, or pride.

The lions cut off one zebra from the rest. The zebra gallops, but a lion draws closer. The lion must catch its **prey** quickly. It can only keep up the burst of speed for a short period of time.

The lion leaps onto the zebra's back. It tries to pull the zebra to the ground. Another lion pounces. Together, they bring down their prey. More lions arrive to feast. The largest lions can eat as much as 88 pounds (40 kg) at a time. Afterward, the lions will rest until they need to hunt again.

Baobab Trees

Several species of baobab trees grow in savannas. These trees can be found in Africa and Australia. Baobabs have thick trunks and thin branches. Their trunks store water. Baobabs provide food and shelter for many animals. For example, weaver birds nest in their branches. Monkeys eat their fruit. Fruit bats get nectar from baobab flowers.

Savannas around the World

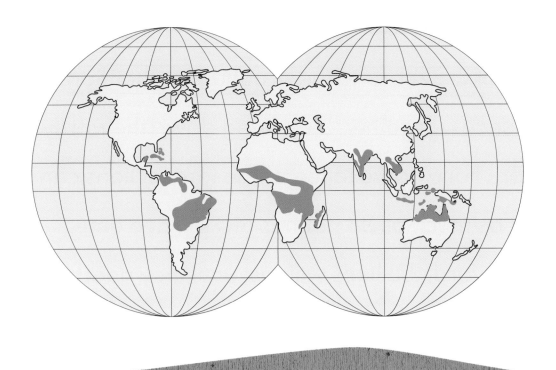

The orange areas of this map show where savannas are located. These biomes are found on several continents where temperatures are warm.

The Savanna Biome

A biome includes the plants, animals, and nonliving things in a particular region. It also includes climate. Trees and grasses are plants that grow on the savanna. Zebras and lions are savanna animals. Water and air are nonliving parts of the savanna.

The savanna biome is a kind of grassland. In addition to grasses and shrubs, scattered trees are part of this biome. Savannas are found in the tropics. They cover parts of South America, Africa, Asia, and Australia. There are about 13 million square miles (33 million sq km) of savannas around the world. They cover about one-fifth of Earth's surface.

Explore Online

Visit the website below. Does it give any new information about African lions that wasn't in Chapter One?

Lion

abdocorelibrary.com/savanna-biomes

Water holes in the savanna attract wildlife.

CHAPTER 2

ABOUT SAVANNAS

Being in the tropics, savannas are warm to hot year-round. Average low temperatures are warmer than 50 degrees Fahrenheit (10°C). Average high temperatures are around 86 degrees Fahrenheit (30°C).

Wildfires limit the spread of forests and maintain savanna biomes.

Rainfall ranges from 8 inches (20 cm) to 157 inches (400 cm) a year.

Savanna biomes usually have a wet and a dry season. Some savannas have long wet seasons that last seven to nine months. Other savannas are very dry. Their dry seasons may last more than seven months.

Trees have trouble growing in dry savannas. But some savannas have enough rainfall to support forests. Factors such as wildfires and poor soil limit tree growth.

A Burning Biome

Wildfires are common on savannas. Dry savanna grasses burn easily. Lightning or human activities can start the fires. The flames get rid of dead plant matter.

Fire as a Tool

For thousands of years, Aboriginal Australians have managed savannas with fire. Flames clear old and dead grasses for easier travel. They allow new grasses to grow. These young and tender grasses feed animals such as kangaroos and wallabies. Aboriginal Australians can hunt these animals for food.

White storks and other animals may hunt at the edge of a wildfire. The flames cause insects and other prey to flee.

Some grasses sprout from seeds after a fire. Other grasses survive because their roots are protected underground. Fires clear out dead grasses that can make it difficult for other plants to grow.

Savanna trees have **adapted** against fire. Some have thick bark that protects their trunks from the flames. Other trees can still grow even if part of the tree dies. However, wildfires sometimes kill trees, especially young ones. They limit the growth of trees in the savanna.

Further Evidence

Look at the website below. Does it give any new evidence to support Chapter Two?

Savanna

abdocorelibrary.com/savanna-biomes

Baobab trees have thick trunks that help them store water.

CHAPTER 3

ANIMALS AND PLANTS IN SAVANNAS

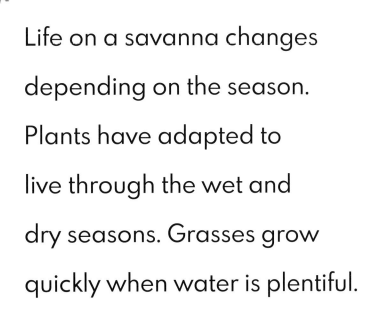

Life on a savanna changes depending on the season. Plants have adapted to live through the wet and dry seasons. Grasses grow quickly when water is plentiful.

When the weather becomes dry, the grass turns brown. The plants stop growing to save water.

Savanna trees have adapted ways to save water. Acacia trees have small leaves that hold in water. Some trees store water in their roots. Trees such as the red bushwillow lose their leaves during the dry season. When the wet season returns, the leaves grow back.

During the dry season, there are fewer plants for animals to eat. Plant-eaters such as zebras and wildebeests **migrate**. They go to areas with more rain and food.

Savanna Interactions

Savanna plants provide animals with food and shelter. Many savanna animals feed on

Wildebeests may migrate as far as 1,000 miles (1,610 km) each year.

grasses, shrubs, and trees. Animals, including kangaroos, rest in the shade of trees. Birds such as yellow-browed tyrants nest in trees. So do tree rats.

Vervets eat mainly plants and fruit. They help spread seeds in savannas.

Animals help plants grow by spreading seeds. Antelope and black-backed jackals eat fruit from trees and shrubs. Seeds from the fruit are released in the animals' waste. Other seeds stick to animal fur and are moved to new areas of the savanna.

Animals also limit plant growth. They can keep trees from taking over the savanna. Large animals may step on or eat tree seedlings. African elephants may crush full-grown trees and shrubs.

Termite Mounds

Termites are insects that live in large social groups. Some species build mounds on the African savanna. They loosen the soil, causing it to hold more water. Their waste has many materials that make the soil healthy. Lots of plants grow around termite mounds. The plants provide food for animals such as geckos and zebras.

Cheetahs are one of the top predators on the savanna. They can run at speeds of up to 75 miles per hour (121 km/h).

Predators affect the amount of prey living on savannas, which then influences plant life. Many prey animals eat plants. These animals change the balance of trees and grass in the biome. If predators eat too many prey animals, certain plants may take over. If there aren't enough predators, the prey may grow in number and strip the savanna of food.

Humans and Savannas

Humans are changing savannas in many ways. People kill large animals such as wildebeests for meat. Humans also kill predators. For example, farmers sometimes kill cheetahs to protect livestock.

People also turn savannas into croplands. But over time, farming can make the soil poor. This makes it difficult for new plants to grow. Too much livestock results in grasses being eaten down to the bare ground. Roads can destroy plants and lead to soil washing away. The area may become desert.

People also work to protect the savanna. Governments and organizations protect some areas of savannas. One example is Serengeti National Park in Tanzania. It covers 5,700 square miles (14,763 sq km) of savanna and acacia woodland. Some people also work to protect savanna animals, such as elephants and rhinoceroses.

People can go on a safari and watch wildlife in the
Masai Mara National Reserve in Kenya.

A giraffe's long neck allows it to eat from high branches of acacia trees and other plants.

Savannas are home to thousands of plant and animal species. National parks on savannas allow people to enjoy wildlife. It is important to protect the plants and animals that make up savanna biomes.

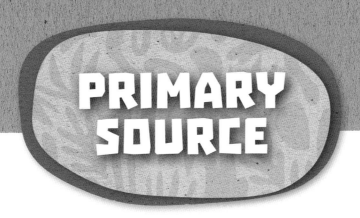

Scientist Liana Zanette talked about how prey species reacted when they heard predators on the African savanna:

> Animals feared lions most. So the king of beasts is definitely the king of beasts. And that was followed by wild dogs, [and] that was followed by cheetahs.

Source: Sam Zlotnik. "How Animals Survive in a Savanna Full of Predators." *Smithsonian Magazine*, 6 Sept. 2022, smithsonianmag.com. Accessed 2 Dec. 2022.

What's the Big Idea?

Read this quote carefully. What is its main idea? Explain how the main idea is supported by details.

EXPLORE THE BIOME

acacia

ostrich

grass

lion

Many plants and animals help make up savanna biomes. Can you identify which parts of this savanna biome are plants? Which are animals? Can you find any nonliving parts of the biome?

giraffe

zebra

rock

Index

Aboriginal Australians, 13
acacia trees, 18, 24

baobab trees, 7

dry seasons, 12, 17–18

elephants, 21, 24

farming, 23–24
fruit, 7, 20

grasses, 5–6, 8–9, 13–15,
  17–19, 23–24

kangaroos, 13, 19

lions, 6–9, 27

seeds, 14, 20
Serengeti National Park, 24

tropics, 9, 11

wet seasons, 12, 17–18
wildebeests, 18, 23
wildfires, 13–15

zebras, 5–8, 18, 21

About the Author

Clara MacCarald is a freelance writer with a master's degree in ecology and natural resources. She lives with her family in an off-grid house nestled in the forests of central New York. When not parenting her daughter, she spends her time writing nonfiction books for kids.